IS SCIENCE RELIGION?

BY

GEOFFREYSON KHAMALA

Published by:

Copyright © 2014 Geoffreyson Khamala

All rights reserved.

ISBN-13: 978-1503333543

ISBN-10: 150333354X

DEDICATION

I dedicate this book to all those who would rather be non-categorized religiouswise.

TABLE OF CONTENTS

DEDICATION iii
TABLE OF CONTENTS iv

LIST OF ACRONYMS & ABBREVIATIONS v
PUBLICATIONS BY GEOFFREYSON KHAMALA vi
ABSTRACT vii

CHAPTER ONE 1

COMPATIBILITY OF SCIENCE AND RELIGION 1

    INTRODUCTION 1
    THE PURPOSE OF KNOWLEDGE 1
    IS RELIGION AND SCIENCE MUTUALLY EXCLUSIVE? 4
    CONCLUSION 18
CHAPTER TWO 19

THE NATURE OF REALITY 19

    INTRODUCTION 19
    RANDOMNESS AND MIRACLES 19
    NON-LOCALITY AND ENTANGLEMENT 25
    PARTICLES, FORCES AND THE HEAVENS 28
    CONNECTEDNESS AND THE NATURE OF REALITY 39
    CONCLUSION 57
CHAPTER FOUR 58

IS SCIENCE A MORAL SYSTEM? 58

    INTRODUCTION 58
    ETHICS OF SCIENCE 58
    THE FUTURE OF SCIENCE AND RELIGION 62
    CONCLUSION 68
CHAPTER FOUR 69

CONCLUSION 69

BIBLIOGRAPHY 70

LIST OF ACRONYMS & ABBREVIATIONS

BCE   -   Before Common Era

CE   -   Common Era

etc.   -   et cetera

i.e.   -   that is

NASA   -   National Aeronautics and Space Administration

Org   -   Organization

WMAP   -   Wilkinson Microwave Anisotropy Probe

PUBLICATIONS BY GEOFFREYSON KHAMALA

1. The Perfect Theory: A Complete Unified Description of the Universe (2014)

2. What is science? Science as an Adaptive Capacity (2014)

3. Is Science Religion? (2014)

4. Wither Globalization Enter Connectedness (2014)

5. The Ultimate Theory: The Perfect Description of the Universe (2015)

6. Tajiriba Spaces: The Solution to Sub-Optimal Outcomes (2015)

7. Zero Unemployment in Kenya: The Utility of Tajiriba Spaces (2015)

8. Reclaiming the Sahara: A Case for Universal Connectedness (2015)

ABSTRACT

We are used to questioning if religion is science. However, it makes more sense to ask ourselves if science is religion. For many people, science and religion seem to be at odds with each other. Conversely, science and religion actually inform each other.

This piece of writing supports the compatibility of science and religion. Science and religion are united by a common purpose but are fundamentally separate methodologically.

Reciprocally, science and religion seek to grasp why the universe exists, the purpose of our existence and what will come afterward.

From this perspective, science and religion are epistemological equivalents. Science and religion are ways of knowing and understanding the universe in order to sustain life.

Science has a lot in common with religion when it comes to critically thinking about the prospects for timeless existence.

The purpose of science is to improve the way humans and other living lives harnesses nature to facilitate continued endurance. Religiosity is meaningless if it does not assist to uphold life.

Keywords: Science, religiosity life, mourning, shared purpose

CHAPTER ONE

COMPATIBILITY[1] OF SCIENCE AND RELIGION

INTRODUCTION

This chapter scrutinizes the purpose of knowledge in order to establish whether science and religion are incompatible.

THE PURPOSE OF KNOWLEDGE

Knowledge is the world's most priceless resource. Science and religion are ways of knowing and understanding the universe. Scientific knowledge comes from experience (observation and experiment) and mental activity (thought, contemplation and reflection).

Religion is the human enterprise to gain knowledge beyond the empirical. The pioneers of the major world religions such as Buddha, Muhammad, Confucius, Moses and Jesus all identified themselves as teachers or prophets.

---

[1] Compatibility is the state of two or more things/situations being able to exist or work together in combination without problems or conflict.

Academic science differs from religion because of its dependence on empirical evidence and testable explanations. Academic science is a deliberate (though random) and rational process geared at studying chance events to quantify doubt and control uncertainty. This explains why scientific results are provisional, always subject to being modified or even overturned by subsequent events. But why do we want to understand the universe?

Knowledge is not an end it itself. Knowledge must have a purpose. Knowledge for its own sake is an exercise in futility. Ways of knowing and understanding (science and religion) hopes to fulfill a goal. If the principle of the universe is to sustain life then the imperative of knowing is to sustain life. Knowledge facilitates adaptation. Science, in the holistic sense, is the adaptive capacity to sustain life (Khamala, 2014b). This is the simplest and all-encompassing definition of science. We seek to know so that we can sustain life.

Since the dawn of humanity, human beings have desired to avoid demise. This truth is evident in mystical and religious pursuits, art, literature, fables, myths, medieval alchemists and contemporary science.

Science is deliberate. Science is a rational process. Science has been successful because it is open to new discoveries. Scientific advances have allowed humans to profoundly alter their environment. Humans have tamed wood, stone, bronze, iron, steel, lumber and even living vegetation in order to sustain life.

But is science a belief? Differently put, is science a religion? For some, science is just another belief system. Indeed, science and religion are mental constructs sustained by human beliefs. For the most part, however, supernatural or metaphysical explanations are outside the scope of academic science.

Religion arose as an attempt to grasp and explain an apparently mysterious and unpredictable world (randomness of nature). Overtime science has become the standard way of explaining and understanding the universe in terms of natural laws. Yet religion did not die as earlier predicted. Why?

Why is religion such a persistent and pervasive feature of society? The reason is simple – the desire to escape death. Death is the reason as to why religion persists. The fear of death is a near-universal human experience and religious beliefs are attempts to ward off this anxiety. Religion is based on the human experience of death. Religion remains a durable feature of human society because of the human aversion to death. For religion, death is not a terminal event.

IS RELIGION AND SCIENCE MUTUALLY EXCLUSIVE?

Religion is an intellectual phenomenon that comprises social institutions, traditions, ritual practices, beliefs, storylines, sacred texts and stories, and sacred places of worship that identify and convey ways and forms of understanding (knowledge) about the universe.

Religion has existed for many thousands of years in every society. Creation myths can be found among the Sumerians, the Babylonians, the Greeks, the Aztecs, the Chinese, the Indians, the Egyptians, the Aksumites, the Nubians and every other early and contemporary community.

The early origin of science is connected with prehistoric religion. It took many centuries for science to emerge as a distinct and organized human activity.

Some religion belief systems acknowledge some superhuman power or powers as in a god or gods. Theism is the belief in the existence of a transcendent deity (monotheism) or deities (polytheism) and their relationship to the universe. Certain godly attributes such as all-powerful (omnipotent), all-knowing (omniscient), unchanging (immutable), perfect and eternal arose from the fusion of Jewish, Greek and Christian influences. The classical conception of the deity (deities) is also found in Judaism, Islam, Sikhism and some forms of Hinduism.

Arguments for the existence of God come in many different forms; some draw on history, some on science, some on personal experience, and some on philosophy.

The primary philosophical arguments in support of theism include the ontological argument (an argument by St Anselm for the existence of a perfect being); the first cause argument (an argument for the existence of an eternal Creator that transcends time, that has neither beginning nor end); the argument from design (a teleological argument for the existence of Creator with a special interest in humanity from the fact the universe is ordered); the moral argument (an argument for the existence of a moral authority that is greater than any of us and that rules over all creation); and miracles (Swinburne, 1977; 2004; Mackie, 1982).

Rival positions to theism that affirm a disbelief in the existence of God (gods) include atheism, agnosticism, pantheism deism, and non-theistic religions like Buddhism.

Argument for atheism takes two forms: a priori and a posteriori. The problem of evil is a posteriori argument denying God's existence (Lactantius, 1871; Russell, 1927).

One of the most common a priori arguments for atheism is the view that religious belief is mere wishful thinking attractive to those who are unable to deal with the reality of life without taking recourse to supernatural phenomenon (Marx, 1844). Karl Marx went ahead to suggest that religion is the opiate of the masses.

Thomas Paine in the seminal essay "Age of Reason" though intimating that he was a firm believer in the existence of God; was skeptical and dismissive of religion as a whole.

Mackie (1982) in *The Miracle of Theism: Arguments for and against the Existence of God* concludes that the evidence in support of the existence of God is not sufficient.

Pantheism is the belief that the universe is synonymous with God (Spinoza, 1667). God is the sum total of all there is and that the combined substances, forces, and natural laws which we see around us are but manifestations of God.

Scholars have argued for pantheism from observations of the unity of nature. Pantheism is also associated with the Egyptian religion when Ra, Isis, and Osiris were identified with all existence.

The pantheism of Hinduism suggests that the impersonal source of all existence is Brahman. The separation of everything into different objects and persons is but a mere illusion - the true reality is the spiritual, incorporeal, and impersonal reality of Brahman, a reality that we can really know nothing about.

Many philosophers through the 17th, 18th, and 19th centuries adopted pantheism in some form, including Spinoza, Hegel, Fichte, and Schelling.

Spinoza (1660) wrote that God and Nature were but two words for the exact same thing and that nothing could possibly exist outside of that single, unlimited substance. The sentiment of pantheism has had a powerful influenced the thoughts and works of poets, philosophers, mystics, and extremely spiritual people. Many ancient leaders, such as the Caesar, claimed divine origin.

In the same vein, paganism represents a wide variety of traditions that emphasize reverence for nature.

Today, as always, the relationship between science and religion evoke passionate disagreements among scholars. Subsequently, scholars have failed to offer a comprehensive explanation of the origin, nature and functioning of religious phenomena.

Evans-Pritchard (1965) and Clifford Geertz (1973) suggested that religion is always embedded within a culture. Tylor & Frazer (1971) demonstrated that religion was suitable to people in relative ignorance but unacceptable in an age of reason and science. August Comte (1891) opined that religion is a stage in human evolution. Humans have evolved through stages: theological, metaphysical and finally positivistic or scientific. Religion is, therefore, a primitive form of experience. Religion is to be replaced with science and its materialistic worldview and empirical means for verification of hypotheses. Similarly, Herbert Spencer (1887) thought that religion is a phase of human evolution. Religion represents an early and primitive phase of human evolution. For him, religion arose out of a form of ancestor worship (ancient beliefs and practices).

Sigmund Freud (2001) rejected religion as a remnant of an age of lack of knowledge and psychosomatic immaturity. For Freud, religion represents a form of neurosis and an essential misrepresentation of reality. He demonstrated how all religions arise out of Oedipal complex especially the need to have restrictions on basic physical drives. Religion provides support for the formation and functioning of the moral sphere or aspect of human nature.

Carl Jung, Freud's successor, held that religion is born out of the collective unconsciousness in which all humans participate and experience to maintain mental balance or health.

Karl Marx rejected religion as the opium of massed individuals (Marx, 1844). He believed religion is fundamentally hypocritical, makes people act like slaves and to accept the status quo. For him, religious worship is a mirage that that hides underlying realities. Although it might profess valuable principles, it ends up siding with the economic oppressors. Religion makes the poor to belief that it is all right that they are poor now, because they will find true happiness in the next life.

Emil Durkheim (1915) rejected the notion that religion is evil or unhealthy to social development. For him, the origin of religions was in the group or the society and give people a sense of belonging. He demonstrated how religion found its earliest and most basic form of representation in the symbol of the totem often in the form of a deity representing the tribe or people as a whole. Religious belief systems offer members of a group a means by which they can identify with one as an organized group. Durkheim predicted that as humans come to identify with the entire species, religion would be transformed into a more rational and scientific approach toward answering questions and organizing common life.

Unquestionably there is incongruity between religion and science as an intellectual activity. However, science and religion are not mutually separate. Both science and religion share historical traditions and reciprocated engagements that exhibit development over time. It is the developments in religious thought that eventually gave birth to science as we know it today.

According to Robert Merton (2001) and Max Weber (2002), it was developments in Protestantism that led to the emergence in seventeenth-century of a culture of individualistic rationalism conducive to scientific modes of thought. In fact, religious beliefs and opinions are not mere superstitions and imaginations. Most religious teachings include scientific claims. However, all this said, science and religion pursue knowledge using different methodologies.

Whereas academic science relies on reason and empiricism, religion acknowledges revelation, faith and sacredness. Today academic science is defined as that body of knowledge about human and the universe that is based on observation, experiment, and measurement. Academic science as the source of understanding of the natural world is based on reason. On the other hand, religion embodies teachings that are based on faith.

Thomas Aquinas (1224 - 1274) in his monumental *Summa Theologica* argued that both human reason and divine revelation were gifts of God. He concluded that theology is superior way of knowing as it is guided by divine knowledge. On the contrary, Nicholas of Cusa (1400-1464) demonstrated the limits of reason by showing how rational analyses of the heavens frequently lead to contradictory results.

The tendency to separate science from religion developed first in antiquity, persisted in the medieval period, reached its climax in the late nineteenth century, and still survives today.

Contemporary science could not emerge until intellectual inquiry was freed from the dogmatic constraints of an ecclesiastically imposed theology. The conflict between emergent science and religion became apparent in the sixteenth and seventeenth centuries as epitomized by Galileo's epic struggle with the Catholic Church over the Copernican system.

Galileo was influential in developing the rational scientific method by his refusal to accept without question statements that were not based on direct evidence but that merely derived their authority from others. Galileo was summoned to Rome immediately to answer for his indiscretion. Tried by the Inquisition and found guilty of heresy, Galileo was forced to recant his Copernican belief.

Scientists suggest that faith as a basis of belief is inconsistent with the scientific conception of steadfast knowledge.

Richard Dawkins, for example, is critical of religion. In *The Selfish Gene* (1976), he popularizes the gene-centered view of evolution. In *The Blind Watchmaker* (1986) he argues against the watchmaker analogy, an argument for the existence of a God based on the complexity of living organisms. In *The God*

Delusion (2006), he asserts that a supernatural creator almost certainly does not exist and that religious faith is a false belief.

Even then, Peter Higgs, an atheist, contends that science and religion are not mutually exclusive. However, as science increasingly facilitates our understanding of the world, more and more people become less motivated to become believers.

This write-up affirms that science and religion are intellectual frameworks which come in handy in the human attempt not only to understand the world but to perpetuate living life indefinitely.

Science as a thought process and as a practice seeks to discover and manipulate the laws of the natural world to protract life (Khamala, 2014b).

Science is supportive of theism and other forms of religion. Religion is edified on prayer, faith, salvation and sanctification while science is a rational process based on evidence.

Science and religion are united by the why question (the purpose) but separated by the how question (the method). Science and religion are adaptive capacities.

Fig. 1: Compatibility of Science and Religion

CONCLUSION

Since the purpose of knowledge is to sustain life, science and religion are not mutually exclusive. Science and religion differ in their approaches but share a common purpose. Science and religion are compatible intellectual processes of attaining knowledge of the natural world in order to perpetuate life.

CHAPTER TWO

THE NATURE OF REALITY

INTRODUCTION

The fundamental nature of reality is one of the unresolved mysteries of the universe. From early on, humans have struggled to understand the natural world. Spiritual and early scientific traditions shape our present-day understanding of the world. This chapter looks into the similarities and differences of religious and scientific approaches to grasping the universe.

RANDOMNESS AND MIRACLES

The occurrence of events is absolutely random. Randomness[2] is the lack of pattern or predictability in events. Randomness may also imply the occurrence of events whose outcomes are not certain.

---

[2] Randomness refers to odds and chance; the lack of pattern, predictability or deliberate purpose in events.

Heisenberg's Indeterminacy Principle demonstrates that the behavior of particles of matter (atoms) is uncertain. This observation is often generalized to apply to humanity as a whole to demonstrate the unpredictability of human behavior and action (and the occurrence of physical events).

Randomness is the result of interactions that may be unknown but exist. This is one of the foundations of religion, but also a motivation for discovery in science and mathematics. Contemporary statistics, calculus, algebra, topology and quantum mechanics are products of the attempts to regularize odds and chance (Hald, 2003).

Prehistoric people supposed that their fate was the outcome of chance and randomness (Johnston, 2003; 2007). They threw dice to ascertain their fate, and this later progressed into games of chance. Most ancients adopted various methods of divination, fortune-telling and prophecy in desperate attempts to circumvent uncertain fate. If and when successful such feats were treated as miraculous events.

A miracle is any improbable event as per known laws of nature or a phenomenon that is not fully explicable by natural causes alone. Because miraculous events are statistically unlikely to happen, their occurrence is often attributed to divine intervention or an unknown outside force (Hume, 1748).

Miracles capture events involving the supposed intervention of supernatural entities in the life of human beings.

Jesus' virgin birth and resurrection are miraculous events that are central to the Christian faith. The gospels also record Jesus performing exorcisms, cures and nature wonders. For example, John 11 records that Jesus raised Lazarus from the dead. Jesus also walked on water.

Other Biblical descriptions of miracles include the parting of the Red Sea, Elijah raising a widow's dead son, Elisha multiplying the poor widow's jar of oil and bringing back to life the son of the well-to-do woman of Shunem.

The Islamic faith treats the Qur'an, Prophet Mohammed and the sacred history as miraculous events. The Arabic word for miracle is *mu'jizah*.

Historically, there have been miraculous claims of the capacity for rainmaking, making barren women to become pregnant, faith healing, etc.

Miracles do not make scientific sense. Scientists point out that religious belief in miracles is inconsistent with the scientific conception of natural law. A natural law is a generalization based on evidence of regular repeatable observation of natural phenomena.

The scientific discoveries made by Nicholas Copernicus, Galileo and Newton during the scientific revolution of the sixteenth and seventeenth centuries began to describe and explain the natural and physical laws by which the earth operates. These discoveries drastically changed the way the world and nature are understood.

Benedict Spinoza (1632-1677) argued that miracles are contrary to universal scientific observation and repetition and therefore absolutely improbable.

David Hume (1711-1776) observed that human experience is uniform (nature is regular with no credible exception) and, thus, was opposed to the claim for highly irregular events such as miracles. He thought of miracles as an interruption or violation of the laws of nature.

Immanuel Kant (1724-1804), a transcendental idealist, in *Religion within the Bounds of Bare Reason* (1793) observed that miracles are only significant in the rise and spread of a religion after that they become redundant. For him, miracles are theoretically possible but they are practically impossible (never occur).

For some scientists, stories of miracles are simply the use of figure of speech, allegory, and exegesis (Cadwallader, 2011).

Nature is mathematical (is characterized by recurring patterns) and that is why mathematics is a useful tool for describing the universe. However, the suggestion that regularities (recurring patterns) rather than exceptions (irregular and singular occurrences) are the basis of scientific understanding is not watertight as it seems.

Simply because miracles are not subject to repetition does not mean they do not occur.

Evidence abounds showing that there are viable scientific explanations for singular unusual events. For example, the Big Bang theory that explains the origin of the universe occurred only once and it has not been repeated. Another singularity is the origin of life on earth.

Science (for example quantum mechanics) describes the world in terms of probabilities rather than definite outcomes. Certain observations cannot be predicted absolutely. Instead, there is a range of possible observations each with a different probability.

NON-LOCALITY AND ENTANGLEMENT

Almost all of physics[3] is based on Einstein's principle of locality. The principle of local action holds that an object is directly influenced only by its immediate surroundings. Distant objects cannot have direct influence on one another. This view ultimately rests upon the assumption that the material universe alone exists. However, quantum mechanics allows for both randomness and the non-locality of entangled particles regardless of how far apart they are.

Non-locality suggests that even when material objects are separated by large distances (potentially even billions of light

---

[3] Physics is the study of nature.

years) they are actually potentially connected in a close and instantaneous manner.

Non-locality occurs due to the phenomenon of entanglement, whereby particles that interact with each other effectively lose their individuality and in many ways behave as a single entity.

In the quantum world objects in "separate" parts of the universe communicate instantaneously violating Einstein's suggestion that the speed of light is the maximum speed for anything in the universe. The implication is that the entire universe is organized in such a way as to anticipate the future thereby violating the laws of causality.

Einstein was so upset by the phenomena of "spooky interaction at a distance" that he declared that the whole of quantum theory was incorrect.

Non-locality is now widely accepted by physicists.

For some scientists, the laws of nature are merely the customary sequence of apparent causes. The principle of causality is not a self-evident truth, nor is it a universal and necessary rather it is based upon long and careful observation of facts. Unusual events like miracles may therefore occur.

The surprise aroused by the occurrence of a miracle is due to the fact that its cause is hidden or unknown (but exists) and the expected outcome is other than what actually takes place. The implication is that if scientists knew the forces (causes) there would be no miracles.

Because of non-locality and entanglement, there are disagreements on whether the natural world (space) is itself a material object, a relationship between material objects or part of a conceptual framework.

PARTICLES, FORCES AND THE HEAVENS

According to standard theory of particle physics, the universe is made up of particles (quanta) and the forces (fields) that keep those objects together. All matter is made up of small particles or atoms.

Particles (tiny bits of matter) are the building blocks of the universe. The number of electrons, protons and electrons determine the properties of those atoms (i.e. elements, molecules or material objects). An element is matter that is composed of the same kind of atoms. A molecule is a group of two or more atoms. Molecules form material objects or matter. Matter is anything that has mass and occupies space. Matter can be solid, liquid, gas or plasma (Boundless, 2014). Matter can undergo a change of state with a contemporaneous energy change.

To complete a missing link in the widely-accepted Standard Model of particle physics, Peter Higgs predicted the discovery of a new kind of elementary particle (Higgs boson) which acts on others to give them mass. For Higgs, particles acquired the property a fraction of a second after the birth of the universe after interacting with a theoretical field (now called the Higgs mechanism).

Experiments in the Large Hadron Collider at Cern in 2012 appeared to validate his predictions. Higgs is an atheist and is uncomfortable with the fact that the Higgs boson is called the 'God particle' as he believes the phrase might upset inflexible believers.

According to standard theory of particle physics, all interactions within the universe arise due to only one of four fundamental forces: gravity, electromagnetism, strong nuclear forces and weak nuclear forces.

The standard model of particle physics (quantum theory) explains all the fundamental forces (electromagnetism, strong nuclear forces and weak nuclear forces) except gravity. Gravity is explained by Einstein's theory of General Relativity.

The theory of General Relativity and the standard model are known to be incompatible with each other. Whereas the quantum theory describes the micro world (atomic and subatomic processes), the general theory of relativity describes the macro world.

In most ordinary physical situations, gravity applies to super-galactic levels of the universe while quantum mechanics describes the smallest structures in the universe. However, some physical situations require an understanding of both the macro and micro structures of the universe, for instance, the state of the universe just before the big bang (the process of the birth of the universe) and a region of space where gravity is so intense that no matter, radiation or light can escape (a black hole).

Perhaps even more intriguing is that from observations of distant supernovae by Saul Perlmutter, Brian Schmidt and Adam Riess, the universe is not just expanding but actually accelerating in its expansion (Nobelprize.org, 2011). The explanation for this scenario is that dark energy (antimatter) acts as a repulsive force over the large scale thereby overcoming gravity.

According to expansion theory, the density of the attractive gravitational pull of matter and the repulsive gravitational push of dark energy are not constants. The density of matter decreases as the universe expands because the volume of space between galaxies increases (not galaxies themselves).

Dark energy and cosmic acceleration are a failure of General Relativity on very large scales. The breakdown of General Relativity and the Standard Model is the driver behind the quest to explain every single aspect of the universe in one theory.

A complete understanding of the universe requires a single model explaining all the fundamental interactions of nature: gravitation, strong interaction, weak interaction and electromagnetism. In addition, the model has to capture dark matter and dark energy.

Candidate theories of include string theory, superstring theory, M-theory and loop quantum gravity. The major predicaments for these candidate theories of everything are that they do not restrict the characteristics of their prediction and/or are inconsistent with observations. More importantly, an understanding of the physical universe can only be achieved with the right knowledge of the elemental nature of reality. Physical space is the illimitable three-dimensional extent in which objects exist and events occur.

Curiosity has driven scientists to the brink of controlling the universe. When confronted with a new phenomenon, we are compelled to search for understanding.

Humans occasionally exhibit their curiosity in their quest to understand the nature of reality primarily in the form of mystery, wonder, miracle or supernatural phenomenon.

Heaven(s) has become a common mythological, metaphysical, religious and cosmological term in reference to the separation between the earth and everything above the earth including the sky, the sun, the stars, the moon etc. Human beings have always been keen to know more about the outer reaches of the universe.

Ancient Egyptians believed in the afterlife. For them, heaven was a far-flung physical place somewhere above the Earth, basically in outer space. According to George Allen in *The Book of the Dead or Going Forth by Day: Ideas of the Ancient Egyptians Concerning the Hereafter as Expressed in their Own Terms* (1974), departed souls endure hurdles as they journey to heaven.

According to Masumian (2002), some Indian religions believe in rebirth and ultimately enlightenment (*Nirvana*). Nirvana is characterized by intense peace of mind (liberation from all suffering) after all craving, distaste, and fantasies have been extinguished. It is blending with the *Brahman* (Supreme Being).

In Buddhism, heavens embody illusionary reality (*samsara*). Buddhists focus more on escaping the cycle of rebirth (ignorance and suffering) and reaching the highest happiness, enlightenment (*Nirvana*).

In Islamic faith, the most valued level of heaven is paradise or Eden (*Firdaus*) to which the prophets, martyrs and other pious people will go at the time of their death. Islamic texts describe immortal life (an afterlife) in heaven as happy, without harmful emotions for those who do good deeds. Heaven is described primarily as a place where every wish is immediately fulfilled on request.

The biblical view teaches that in the beginning God created the heavens and the earth (the whole universe) (Genesis 1:1; 2:1; Jeremiah 23:24; Acts 17:24). There were three heavens according to the Jewish tradition from the Bible. The first heaven is earth's immediate sky, or the clouds and atmosphere that surrounds earth; the second is outer space as far as it stretches or where earth's atmosphere ends; and the third is the place where God, angels and spirits of just beings dwell (Deuteronomy 17:3; Jeremiah 8:2; Matthew 24:29; Deuteronomy 10:14; 1 Kings 8:27; Psalms 115:16; 148:4). The third heaven is yet to be explored by scientists because humans are yet to attain the capability observe it via the telescope.

Most academic scientists do not believe in heaven but fancy the possibility of an afterlife. Stephen Hawking, the foremost living scientist, dismisses the belief in the heavens as a mere 'fairy tale' (The Guardian, 2011).

Contemporary science, like religion, is obsessed with outer space, the void that lies further than the peak reaches of Earth's atmosphere linking all other celestial bodies in the universe.

The study of outer space include prestigious disciplines such as space science, astronomy, cosmology, planetary science, aerospace engineering, orbital mechanics, space travel and exploration, astrobiology, astrochemistry, astrophysics, space archeology and remote sensing.

Today the National Aeronautics and Space Administration (NASA), the United States government agency that is responsible for the space program, aeronautics and aerospace research, is the most high-status organization to work for.

The outer space environment is not a perfect vacuum since it contains stray gas molecules made up of plasma (cosmic dust, hydrogen and helium, electromagnetic radiation, magnetic fields and neutrinos) (Genz, 2001). However, the concreteness of these stray gas particles is so low that they can be thought of as almost missing. The gravitational interaction of large bodies in space, such as planets, stars, galaxies and meteoroids, pulls gas molecules close to their surfaces leaving the space between celestial bodies virtually empty (of matter).

Space scientists use radio waves, cosmic microwaves, infrared, visible light, ultraviolet, X-ray wavelengths and Gamma-ray bursts observatories to study objects in space. Ground-based observatories can only observe nearby galaxies.

The space-based observatory like the Wilkinson Microwave Anisotropy Probe (WMAP), designed to measure a background radiation that pervades the universe, provides a snapshot of the cosmos as it existed some 380,000 years after the birth of the universe. This has enabled scientists to not only find lots of worlds around other stars but even to characterize their atmospheres. It is now possible to discuss the composition the atmosphere and the weather on worlds hundreds of light years away. However, the discovery of dark matter and dark energy gives credence to the biblical view that there is unexplored "third heaven".

For believers, heaven is often described in contrast to hell or the underworld. Related to this, the scientific realm is for decades now still grappling with the veracity of the hollow Earth theory. According to this hypothesis, there exist subterranean universe underneath Earth (interior space) (Standish, 2007).

Many people who come close to death and had near death experiences report meeting relatives or entering unearthly dimension, which share similarities with the religious impression of heaven (Lommel, 2011). They report experiencing intense feeling of love, peace, joy, consciousness or a heightened state of awareness beyond human comprehension. There are also reports of distressing experiences which share some similarities with the perception of hell.

CONNECTEDNESS AND THE NATURE OF REALITY

Connectedness shifts our understanding of the nature of time, being and the universe (Khamala, 2014a). According to connectedness, heavens, the idea of God, infinity, Big Bang, black holes, time travel and immortality demonstrate the human desire to sustain life.

Connectedness offers a more believable unified, comprehensive and universal description of the physical world.

Theory	Interaction	Function	Range	Exchange Particle
Standard Model of Particle Physics (Quantum Theory)	Weak	Causes certain forms of radioactive decay	Very short	Bosons
	Electromagnetic	Interactions between charged particles. Holds atoms and molecules together Responsible for light and chemical properties of matter.	Long range (theoretically infinite but limited due to canceling effects of + and - charges and magnetic poles)	Photons
	Strong	Binds nucleons (protons and neutrons) together	Distance of adjacent nucleons in nucleus	Gluons
General Relativity	Gravity	Acts on all objects with mass (and occupy space). Responsible for orbit of Earth around Sun. Binds stars, planets and galaxies together.	Long range (infinite)	Curvature of space-time Graviton (hypothesized)
The Perfect Theory	Local (Nanoscale) and Non-local (large-scale)	Acts on everything in the universe with or without mass. Links physical and non-physical phenomena.	Infinite (short and long) range	Connectedness

Is nature local or nonlocal? Particle physicists (classical mechanics) say nature is local. Einstein (General Relativity) also suggests that the physical world is local.

The notion of locality holds that one particle influences another only by direct contact or via some agent field and in addition that this influence can travel no faster than light. Locality requires a link between subatomic particles.

Non-local action means that there exist interactions between events that are too far apart in space and too close together in time for the events to be connected even by signals moving at the speed of light.

Quantum theorists demonstrate that quantum reality is non-local (Grib & Rodrigues, 1999). Non-locality in quantum mechanics finds expression in entanglement. Entanglement raises the odds for instantaneous communication (including teleportation[4], invisibility cloaks and time travel) (Darling, 2005; Popescu, 2014).

---

[4] Teleportation refers to paranormal phenomena and anomalies involving the transmission of information (matter or energy) from one location to another without traversing the physical space between them.

Almost all religious traditions suggest that the universe is both local (human earthly experience) and nonlocal (heaven, nirvana or enlightenment). Religion is by definition futuristic.

According to the Perfect Theory the nature of space is both local and nonlocal (Khamala, 2014a). The model acknowledges the undivided wholeness inherent in a more complete account of nature.

Our common sense also tells us that the universe is exclusively composed of fundamentally separate things (i.e. atoms, tiny bits of matter, particles or material objects). Besides the common sense view of space is an absolute vacuum. The truth is that space (nature) is actually linked, continuous and full of activity. The interconnections and interactions give rise to connectedness. The universe is connected as a continuous whole.

The universe is made up of spatial (space) and non-spatial extensions. Extension herein refers to the attributes of a phenomenon by which it occupies space. The theory implies that besides the familiar three-dimensional physical world (space), there are twelve (12) major categories of non-spatial dimensions: gravity, electromagnetism, strong nuclear forces, weak nuclear forces, dark matter, dark energy, mental faculties, senses, emotions, time, life and death.

All events and processes that take place in space (in nature) irrespective of the distances that separate them are linked, uninterrupted and related in an intimate and instantaneous manner. This is why in the process of interaction pairs particles effectively lose their individuality and in many ways behave as a single entity.

All objects (including physical events and processes) instantaneously know about each other's state and act in response as a combined system in view of that refuting Einstein's speed of light theorem. For example, telephonic conversations and internet communications are instantaneous from virtually anywhere, anytime.

Instantaneous action holds the key to practical future technologies such as quantum computing, cryptography, gambling, remote sensing, prophecy and the possibility of the afterlife.

Time is simultaneous. The past, present and future are all happening simultaneously or at the same time. From the time of its birth the universe is separate but remains connected by an instantaneous wholeness.

The connectedness of the universe depends on extensions of the universe including gravity. There is gravity everywhere. Gravity is the natural phenomenon by which particles of matter appear to attract every other particle.

Besides gravity, there is dark matter and dark energy. NASA's Wilkinson Microwave Anisotropy Probe (WMAP) revealed that visible matter makes up a mere 4.6 percent of the cosmos.

The rest of the space is far from empty, however. Dark matter accounts for 23.3 percent of the universe while dark energy takes up a massive in 72.1 percent of the universe. Dark matter consists of the unseen particles that keep our universe together.

Dark energy is also present throughout the universe and is the force that works against gravity. Dark energy makes up nearly three quarters of our universe. It is the force responsible for the acceleration of the expansion of the universe at an ever-increasing rate since the Big Bang.

Gravity, dark matter, dark energy, electromagnetism, strong nuclear forces, weak nuclear forces, mental faculties, senses, emotions, time, life and death (or simply connectedness) is responsible for keeping the Earth and the other planets in their orbits around the Sun; for keeping the Moon in its orbit around the Earth; and for various other phenomena observed on Earth and throughout the universe.

Virtually everything in space is in motion. Now, the expectation is that every other constituent part of matter will always travel in the straightest possible line if there are no external forces at work. In view of that, without an external force, two constituent parts of matter travelling along parallel paths will always remain parallel. They will never meet. However, it appears they do meet because the whole universe is connected. In fact, constituent parts of matter that start off on parallel paths sometimes end up colliding. This can occur because of gravity, the predictable force that attract those objects to one another or to a single, third object in the universe.

Isaac Newton demonstrated that each constituent part of matter attracts every other constituent part of matter with a force that is directly proportional to the product of their masses and inversely proportional to the square of the distance between them. Mass is the amount of matter in something. The more massive something is, the more of a gravitational pull it exerts. Gravity also depends on distance. Gravity decreases with distance such that an object far away from a planet or star feels less gravity. Of course, gravity goes hand in hand with dark matter and dark energy, something which Isaac Newton was unaware of.

Albert Einstein explained this phenomenon is a similar fashion thereby supporting the reality that the whole universe is connected. Einstein noted that things in space have inertia, that is, they travel in a straight line unless there is a force that makes them to stop or change. For him, the movement of things in space is influenced by gravity. Einstein suggested that gravitation is a byproduct of the curvature of space-time governing the motion of inertial objects due to the presence of matter. However, he did not state through which mechanism. Einstein was also unaware that most of deep space (the vast stretches of empty area between planets, stars and moons) has abundant (dark) energy. So, Einstein's space-time curvature is simply a good illustration of connectedness.

Why a quantum particle does behave erratically is because not all properties of a quantum particle can be measured with complete accuracy. That is why the Copenhagen interpretation (first posed by Niels Bohr) suggests rightly that a quantum particle doesn't exist in one state or another but in all of its possible states at once. When observed, an object is essentially forced to take one state or another (probability), and that's the state that we observe. The object may be forced into a different observable state each time it is observed. This position concurs with Werner Heisenberg's conclusion that our observations have an effect on the behavior of quanta.

Why is uncertainty an integral component of space? Simple at a tiny scale it is impossible to measure both the spatial position of a physical system and its momentum (mass times velocity) with arbitrary accuracy because the particle and the rest of the universe behave as a single entity. In other words there is no telling which particle was measured first. Therefore, the more precisely the position of a particle is determined, the less precisely the momentum is known in this instant, and vice versa.

It is impossible to measure anything without disturbing it. For instance, any attempt to measure a particle's position must randomly change its speed. Indeed, quantum experiments have demonstrated the seemingly unavoidable tendency of humans to influence the situation and velocity of small particles by simply observing the particles. Even the light experimenters use to help them better see the objects being observed can influence (accelerate) the behavior of quanta changing its velocity and speed. Uncertainty also confirms the reality that the whole universe is linked in an immediate way.

Does the uncertainty about an object's position and velocity makes it difficult for scientists to determine much about the object? Not really. However, these uncertainties or imprecision in assigning exact simultaneous values to the position and momentum of the particle has profound implications for such integral notions as causality and the determination of the future behavior of an atomic particle.

Causality makes our everyday life orderly. Causality is such that events in the present are caused by events in the past and events in the present act as causes for what happens in the future. Simply put, a cause should always precede its effect. These restrictions are consistent with the assumption that causal influences cannot travel faster than the speed of light and/or backward time travel.

According to special relativity, it would take an infinite amount of energy to accelerate a slower-than-light object to the speed of light. Nevertheless, results of tests of Bell's theorem appear to demonstrate that some quantum effects travel faster than light. This is a pointer to the reality that apparently separate particles (and events) can influence each other instantaneously over great distances.

Isaac Newton demonstrated that an action always prompts a reaction. Quantum theory though manifestly non-local, suggests that whatever we can observe (quantum facts) are always local. Accordingly, the possibility of observing a non-local effect is remote (indeed non-existent).

Cleve Backster in *Evidence of a Primary Perception in Plant Life* (1968) established that plants react to trauma in their local environment. The Backster Effect showed plants display emotion in life-threatening situations. Plants are able to read the human mind (non-physical interaction) before the implementation of the threatened act and express their reaction through emotions.

Plants can perceive and measurably respond to intentional human thought and actions. Nevertheless, this theory has been criticized by some scientists who argue that plants lack a nervous system and can, therefore, not sense and respond to external stimuli. However, recent research has shown that plants can respond to electrical impulses (and radiation) thereby to a degree confirming Backster's experiments on plant psychology.

Plant perception does not deviate from our everyday view of reality. Living things do communicate with other life-forms without physical contact and that communication is immediate. Chances are that plants, animals, micro-organisms, planets, stars, other terrestrial objects and all other phenomena in the universe are living things. The universe (biotic and abiotic entities) is alive akin a super-organism. So, each component of the universe acts on the universe instantaneously all the time for eternity.

The universe as a whole reacts instantaneously but the effects are observed gradually over time. The first significant action was the Big Bang giving birth to science (and religion). Science is the practical and intellectual adaptive activity capacity to preserve life (endlessly in the long term).

Quantum mechanics doubt if non-local effects will ever be observed. Local and non-local effects are what we call experience (trial and error). We observe (or rather experience) them every day. This reality informs the debate on nature versus nurture in the case of human development.

Supporters of the nurture position believe that at birth humans are essentially a blank slate, and that their environment as they grow and develop is the only factor that determines characteristics of the individual. Thus matters of choice of profession, mate, musical preferences, morality, etc. are determined by society. Believers in the nature position, on the other hand, say the genetics is crucial in development, and that the characteristics of an individual are determined at birth. Some religious traditions talk of the original sin (nature) but still give individual human beings the choice (nurture) to determine their destiny.

According to the Perfect Theory, space at its smallest and largest distances is characterized by uncertain relations (randomness) because local and non-local effects are gradual and ongoing. Space is not empty. This reality explains natural events and processes (i.e. rainfall, earthquakes, volcanic eruptions, tides, gravity, etc.) and human events and processes (i.e. schooling, marriage, science, religion, crime, war, poverty, underdevelopment etc.). Simply put, local and non-local effects explain evolution. Biological evolution has produced the diversity of life on Earth.

The brain holds and processes all emotions, thoughts, memories and keeps track of the ongoing functions of the body like breathing pattern, eyelid movement, hunger and the movement of the muscles.

Mental capacities enable humans to understand why and how fields such as electricity, gravity and magnetism that are spread out over vast regions when disturbed could vibrate and travel as waves.

Humans sense and respond to external events. For instance, facts indicate that an observation does shape reality. Microscopes and telescopes are extensions of human's experimental senses.

Judgment of space can also be also altered by physical alterations to the brain such as with traumatic brain injury, psychoactive drugs, temporal illusions, age, hypnosis and neurological diseases such as Parkinson's disease and attention deficit disorder. For example, under hypnosis, some patients allegedly, have vivid memories of past lives. During the ordeal, they are capable to travel to other locations.

Travelling forward in time (travel into the future) and backward time travel appears in ancient folk tales, myths and religious traditions.

According to connectedness, humans have some ability to anticipate the future. Precognition is real. However, knowing the future changes the upcoming since everything in this universe is connected. All effects in the universe are related to all others. When particle attributes (such as position, velocity, polarization, etc.) are observed the rest of them are affected instantaneously even if they are universes apart.

The Perfect Theory is the closest humanity has ever come to having a complete unified description of nature. According to The Perfect Theory, electromagnetic force, the weak force, the strong force, gravity, dark matter, dark energy, mental faculties, senses, emotions, time, life and death are properties of space itself.

The theory can account for phenomena that occur at very small distances in the realm of atoms and subatomic particles, that is, the electromagnetic force (which holds atoms and molecules together), the strong force (which holds protons and neutrons together), and the weak force (which causes certain forms of radioactive decay). The theory also captures the effects of gravity around large masses. The theory explains the formation of the universe, why the moon orbit around the Earth, why humans experience a wide range of emotions and thoughts, why life-forms respond to physical and chemical stimuli and account for many other phenomena in the universe.

CONCLUSION

Heated discussions concerning the character, embodiment and the mode of existence of the natural world date back to prehistory. This chapter has demonstrated that spiritual and scientific traditions complement each other in shaping our present-day understanding of the world.

CHAPTER FOUR

IS SCIENCE A MORAL SYSTEM?

INTRODUCTION

This chapter demonstrates that, like religion, science is a moral system[5].

ETHICS OF SCIENCE

Do morals exist? Without moral systems it is not possible to sustain life. Religion is a moral system. The central tenets of religious faith are about right and wrong. Religion informs what we know today as equal rights. Religion and science reinforce each other. Science is also an ethical system.

Aristotle in *Nicomachean Ethics* declares that whatever leads to greater *eudaimonia*, or happiness, is what is moral.

---

[5] A Moral system is coherent and systematic set of standards, principles, rules, ideals, and values that define virtuous conduct

Non-theistic Niccolò Machiavelli in his classic *The Prince* (1532) exalts power. According to Machiavelli, morality is okay if it accomplishes a political end.

Friedrich Wilhelm Nietzsche (1844 – 1900) in *The Gay Science* (1882) observed that given the development in science and the subsequent secularization of society "God is dead" and the end result is likely to be the demise of any universal perspective on things, and any coherent sense of objective truth.

Immanuel Kant *in Groundwork for the Metaphysics of Morals* (1785) based his deontological moral system on the demands of the categorical imperative: Act only according a moral maxim whereby you can at the same time will that it should become a universal law.

Kant's Categorical Imperative is based on a conception of fairness and universalizability. Kant meant that people should contemplate whether they would desire the moral principle underlying their action to be elevated to a universal law that would govern others in comparable state of affairs. The contemplated action should be avoided if it did not meet the requirements of this maxim.

For Kant, morality is a priori. All duties and obligations derive from pure reason (the capacity to know separate from empirical experiences) rather that practical reason. Immanuel Kant located religion in emotion and held that it was the result of a practical reasoning. For Kant, religion and faith were separated from the realm of reason.

Kant's ethical system neglects to identify and or validate the existence a consistent moral law for everyone. In retrospect, Benjamin Constant wondered whether one must (if asked) reveal to the enquiring murderer the location of his victim. The only way to save the victim is to lie. This explains why Kant in the now infamous essay *On a Supposed Right to Tell*

Lies from Benevolent Motives (1797) replied that it is indeed one's moral duty to be truthful to the would-be murderer.

According to connectedness, the natural world is governed by a universal law. The law of the universe is to sustain life. Life (as in a purposeful life) is an end in itself.

Fig. 2: The Meaning of Life

Science and religion are moral systems united by a common purpose. Science as the moral system is the means to that end. Science and religion are mental frameworks that seek to protect life.

Morality is inborn (nature) but reinforced through experience and learning (nurture). Science and religion are realms of human experience in the quest for knowledge to sustain life. This explains why consciousness is a fundamental aspect of our existence.

THE FUTURE OF SCIENCE AND RELIGION

Science and religion are built on hope. Underlying the tension and conflict between the religious divide are identity-based deep-rooted prejudice and discriminatory attitude. However, the expectation is that religion and other identities (i.e. families, lineages, clans, ethnicities, states, races and civilizations) that bind and partition humanity will give way to universal connectedness. Connectedness is what drives science. Science is a cooperative practical and intellectual activity to sustain life.

Most science is not done in schools, universities, research facilities etc but rather is through trial and error (experience). Science and religion are an awareness of the perils of mortal life and the benefits of immortal living. In a sense it is possible to say science started with religion and will end with religion. The central guiding principle of the universe is the preservation of life.

In the future people will have a choice to avoid some engagements they consider are a violation of the principle of the universe. People may opt out of activities that put their lives at risk or that endanger the lives of others (e.g. the armament industry).

Death is the problem of science and religion. Mourning is the most obvious public expression of human disenchantment with death (Khamala, 2014a). Scientists deploy preventive and curative efforts to prevent, postpone or eliminate death. Scientists also discover, improve and deploy technology to make the lived experience worthwhile. The clergy reinforce the idea that in the event of death there exist possibilities so long as one leads a virtuous existence here on Earth.

Most people (even dogmatic scientists) seek the services of the clergy when on the deathbed for reassurance that their passing away does not end it all. It is even alleged that moment before his death, Albert Einstein, the foremost scientist, agonized over his conviction in the preference of science over religion.

The last words attributed to Archimedes of Syracuse, the greatest mathematician of all time, who died in 212 BCE during the Second Punic War are: "Do not disturb my circles". These rebuke was addressed to a Roman soldier for bothering Archimedes who was then preoccupied with the study of circles in the mathematical illustration he was supposedly drawing. The soldier ignored Archimedes' admonishment and the instructions from the victor General Marcus Claudius Marcellus who had conquered the city of Syracuse after a two-year-long siege.

Morality impacts our everyday decisions, choices and outcomes. Moral systems ensure fair play and harmony in the universe. Living creatures eat, play, reproduce, learn and avoid predators in the struggle for collective existence. Moral systems (particularly thoughts and feelings of guilt, reciprocity, obligation and expectations) help in the balancing of self, selfless and collective interests.

Humans and non-humans survive through fellow feelings (cooperation and mutual assistance) (de Waal, 1997). Pursuance of self-interest regularly attracts denunciation. Likewise, the intentional killing of oneself (self-sacrifice) is an anomaly arising from despair as a result of stress factors (i.e.

predicaments in interpersonal relationships, financial difficulties etc.), substance use, alcoholism and medical condition (i.e. bipolar disorder, schizophrenia, etc).

For long, suicide and attempted suicide have attracted some form of punishment including criminal chastisement. People attempt or commit suicide when they fail to connect with the principle of the universe. Potential and attempted suicide victims seek the services of counselors, psychiatrists and the clergy.

Unlawful killing of others (homicide and manslaughter) is also discouraged. Murder convictions typically draw the harshest sentences of any crime on the understanding that the victim's existence is permanently deprived.

The scientific process is governed by ethical considerations (Hashmi & Lee, 2004; Somerville & Atlas, 2005). Scientific codes of ethics require scientists to reflect on the ethical implications of scientific research and its applications.

Scientists are required to foresee and avoid potential dangerous applications of their work. Equally, no religion allows the killing of (innocent) people. This is basically because science and religion are adaptive capacities.

For long, knowledge acquisition and application has not been morally neutral. Science and religion, as knowledge systems, have been in the spotlight in recent times for endangering lives. Scientists and the clergy have been associated the development of hideous bioweapons, atom and hydrogen bombs, Nazi experiments, the death ray, among other forms of corrupting tendencies. Besides, methodical knowledge regularly put a low value on spiritual knowledge. These realities paint the picture of science and places of worship's interface with wider humanity at the crossroads.

The two-facedness of science (useful and harmful; methodical and spiritual), among other justifications, informs the desire to unite all knowledge in every possible way. The unity of the sciences has to be premised on the purpose of knowledge. The moral of science is to safeguard life.

CONCLUSION

This chapter has made obvious how methodical science and spirituality are moral systems since they are motivated by a common purpose. It emerged that emphasis on the centrality of science in the holistic sense will enhance collective life preservation.

CHAPTER FOUR

CONCLUSION

Science and religion are complimentary spheres of human life. Science and religion are ways of thinking and knowing the world to maintain life. However, scientific progress is often confronted by moral consequences. Science and religion share a common problem (i.e. death) and aspire for a solution (i.e. perpetual life).

People who make fun of religion or misuse it have not properly grasped the essence of religious conviction.

Religion provides the vision and science attempts to fulfill the vision. Science and religion are united by a common purpose but are fundamentally separate methodologically.

Science and religion complement each other in the search for purposeful life.

BIBLIOGRAPHY

Allen, Thomas George (1974). "The Book of the Dead or Going Forth by Day: Ideas of the Ancient Egyptians Concerning the Hereafter as Expressed in Their Own Terms". *SAOC* vol. 37; University of Chicago Press.

Aristotle (2002). *Nicomachean Ethics*. Trans. Joe Sachs. Focus Philosophical Library: Pullins Press.

Backster Cleve (1968). "Evidence of a Primary Perception in Plant Life". *International Journal of Parapsychology* vol. 10, no. 4, pp. 329-. 348.

Bell Stewart John (1987). *Speakable and Unspeakable in Quantum Mechanics*. Cambridge: Cambridge University Press.

Boundless (03 July, 2014). "Phases of Matter." *Boundless Physics*. Boundless. Retrieved Tuesday, November 18, 2014 from https://www.boundless.com/physics/textbooks/boundless-physics-textbook/fluids-10/introduction-91/phases-of-matter-335-6279/

Cadwallader Alan (eds.) (2011). *Hermeneutics and the Authority of Scripture: Task of Theology Today*. ATF Press.

Comte Auguste (2009) [1891]. *The Catechism of Positive Religion*. Cambridge University Press.

Darling David (2005). *Teleportation: The Impossible Leap*. Wiley.

Dawkins Richard (1976). *The Selfish Gene*. Oxford: Oxford University Press.

Dawkins Richard (1986). *The Blind Watchmaker*. New York: W. W. Norton & Company.

Dawkins Richard (2006). *The God Delusion*. Boston: Houghton Mifflin.

de Waal Fans (1997). *Good Natured: The Origins of Right and Wrong in Humans and Other Animals*. Cambridge, MA: Harvard University Press.

Durkheim Emile (1995) [1915]. *The Elementary Forms of the Religious Life*. Trans. Karen E. Fields. The Free Press.

Evans-Pritchard, E. E. (1965). *Theories of Primitive Religion*. Oxford University Press.

Freud Sigmund (2001). *The Complete Psychological works*. Ed. James Strachey. Vintage Paperbacks.

Geertz Clifford James (2000) [1973]. *The Interpretation of Cultures*. Basic Books.

Genz Henning (2001). *Nothingness: the Science of Empty Space*. Basic Books.

Grib Andrey Anatoljevich & Rodrigues Waldyr Alves (1999). "Nonlocality in Quantum Physic", Springer, July 31, 1999.

Hald Anders (2003). *A History of Probability and Statistics and Their Applications before 1750*. Wiley-Interscience.

Hashmi Sohail & Lee Steven (2004). *Ethics and Weapons of Mass Destruction: Religious and Secular Perspectives*. Cambridge University Press.

Hawking Stephen W. (1988). *A Brief History of Time*. New York: Bantam Books.

http://www.guardian.co.uk/science/2012/dec/26/peter-higgs-richard-dawkins-fundamentalism

Hume David (2000) [1748]. *An Enquiry Concerning Human Understanding*. Ed. Tom L. Beauchamp. New York: Oxford University Press.

Johnston Sarah Iles (ed.) (2003). *Religions of the Ancient World: A Guide*. Harvard University Press.

Johnston Sarah Iles (ed.) (2007). *Ancient Religions*. Belknap Press.

Kant Immanuel (1781) [1848]. *Critique of Pure Reason*. London: William Pickering.

Kant Immanuel (1785) [2005]. *Groundwork for the Metaphysics of Morals*. Trans. Thomas Kingsmill Abbott. Peterborough, Ont.; Orchard Park, NY: Broadview Press.

Kant Immanuel (1793) [2009]. *Religion within the Bounds of Bare Reason*. Trans. Werner S. Pluhar. Indianapolis: Hackett Publishing Company.

Kant Immanuel (1797) [1949]. "On a Supposed Right to Tell Lies from Benevolent Motives". In *Critique of Practical Reason and Other Writings in Moral Philosophy*. Ed. and trans. Lewis White Beck. Chicago: University of Chicago Press. PP. 346-5

Kenny Anthony (1969). *The Five Ways: St. Thomas Aquinas Proofs of God's Existence*. New York: Schocken Books.

Khamala Geoffreyson (2014a). *The Perfect Theory: A Complete Unified Description of the Universe*. Tajiriba Foundation.

Khamala Geoffreyson (2014b). *What is Science! Science as an Adaptive Capacity*. Tajiriba Foundation.

Lactantius (1871). "On the Anger of God. Chapter XIII". *The works of Lactantius*. Rev. Roberts, Alexander; Donaldson, James (eds.). https://archive.org/stream/workslactantius00lactgoog#page/n40/mode/1up. Retrieved Friday, July 04, 2014.

Lommel Pim van (2011). *Consciousness Beyond Life: The Science of the Near-Death Experience.* Harper One.

Machiavelli Niccolò (1532) [2003]. The Prince. Dante University Press.

Mackie John Leslie (1982). *The Miracle of Theism: Arguments for and against the Existence of God.* Oxford University Press.

Marx Karl (1844). "A Contribution to the Critique of Hegel's Philosophy of Right". http://www.marxists.org/archive/marx/works/1843/critique-hpr/intro.htm. Retrieved Friday, July 04, 2014.

Masumian Farnaz (2002*).* *Life After Death: A Study of the Afterlife in World Religions.* Kalimat Pr.

The Guardian (15 May, 2012). "Stephen Hawking: 'There is no heaven; it's a fairy story'". Retrieved Tuesday, November 18, 2014 from http://www.theguardian.com/science/2011/may/15/stephen-hawking-interview-there-is-no-heaven

Merton Robert (1938) [2001]. *Science, Technology and Society in Seventeenth Century England*. New York: Howard Fertig.

Nietzsche Friedrich Wilhelm (1882) [2001]. The Gay Science. Cambridge University Press.

Nobelprize.org (2011). "The Nobel Prize in Physics 2011 Press Realease". Retrieved Wednesday, November 19, 2014 from http://www.nobelprize.org/nobel_prizes/physics/laureates/2011/press.html

Popescu Sandu (2014). "Nonlocality beyond quantum mechanics," *Nature Physics* 10, 264–270.

Potter Karl H., ed. (1970–1995). *Encyclopedia of Indian Philosophies*. 5 vols. Princeton NJ; Delhi: Princeton University Press; Motilal Banarsidass.

Russell Bertrand (1927). "Why I Am Not a Christian". http://www.positiveatheism.org/hist/russell0.htm. Retrieved Sunday, November 16, 2014.

Somerville Margaret & Atlas Ronald M.(2005). "Ethics: A Weapon to Counter Bioterrorism". *Science* Vol. 307 no. 5717 pp. 1881-1882 *DOI:* 10.1126/science.1109279

Spencer Herbert (2020) [1887]. *Herbert Spencer's theory of religion and morality*. Princeton University.

Spinoza Baruch (1667). Ethics. http://capone.mtsu.edu/rbombard/RB/Spinoza/ethica-front.html. Accessed Friday, July 04, 2014.

Standish David (2007). *Hollow Earth: The Long and Curious History of Imagining Strange Lands, Fantastical Creatures, Advanced Civilizations, and Marvelous Machines Below the Earth's Surface*. Da Capo Press.

Swinburne Richard (1993) [1977]. *The Coherence of Theism*. Oxford University Press.

Swinburne Richard (2004). *The Existence of God*. Oxford University Press.

Tylor Edward and Frazer Bunnett (1920) [1871]. Primitive Culture. New York: J.P. Putnam's Sons.

Weber Max (2002). *The Protestant Ethic and the Spirit of Capitalism*. Trans. Peter Baehr and Gordon C. Wells. Penguin Books.

www.ingramcontent.com/pod-product-compliance
Lightning Source LLC
Chambersburg PA
CBHW071753170526
45167CB00003B/1013